QUESTIONS

JOHNS HOPKINS
UNIVERSITY PRESS

AARHUS UNIVERSITY PRESS

Questions

PIA LAURITZEN

QUESTIONS

© Pia Lauritzen
and Johns Hopkins University Press 2023
Layout and cover: Camilla Jørgensen, Trefold
Cover photograph: Poul Ib Henriksen
Publishing editor: Karina Bell Ottosen
Translated from the Danish by David Possen
and Heidi Flegal
Printed by Narayana Press, Denmark
Printed in Denmark 2023

ISBN 978-1-4214-4714-8 (pbk)
ISBN 978-1-4214-4715-5 (ebook)

Library of Congress Control Number: 2022949356

*Special discounts are available for bulk purchases of this
book. For more information, please contact Special Sales at
specialsales@jh.edu.*

Published in the United States by:

Johns Hopkins University Press
2715 North Charles Street
Baltimore, MD 21218
www.press.jhu.edu

Published with the generous support of the
Aarhus University Research Foundation

Purchase in Denmark: ISBN 978-87-7597-294-4

Aarhus University Press
Helsingforsgade 25
8200 Aarhus N
Denmark
www.aarhusuniversitypress.dk

PEER
REVIEWED

MIX
Paper
FSC FSC® C010651

CONTENTS

CALLING QUESTIONS INTO QUESTION?

"MAY I ASK YOU SOMETHING?"

This line opens many a short conversation, and it is also a fitting opening for a short book like this. The question "May I ask you something?" seldom prompts more than a one-word response, and there's a reason for that: It's a question that resolves the issue at hand before even raising it.

The issue at hand is *whether or not we may ask a question*. But in asking whether we may ask or not, we've already started asking. We've already presumed that the answer is "Yes." Which is why we sometimes don't even wait for our question to be answered before moving on to our next question, our real question: "Which way is Main Street?" or "Do you have this in size 10?"

Some philosophers hold that the question "May I ask you something?" isn't really a question at all, since it doesn't permit multiple answers, but instead presupposes one specific answer. Some philosophers hold the opposite: that the nature of questions is to resolve and close the issues they raise. As for the rest of us, we just ask away!

We start the day by asking our loved ones whether they've slept well. At school, the teacher asks who wants to read out loud. At work, an employee asks for a deadline. Meanwhile, a researcher writes an article answering her carefully formulated research questions, while a journalist prepares questions for an interview. In a courtroom, the accused is sentenced to prison after failing to adequately answer the prosecutor's questions. In parliament, the politicians practise the art of persuasively answering every question they get, whatever the topic.

Life is full of questions. Ask a stupid question and you'll get a stupid answer. Still, if you never ask, you'll never learn. And you've got to ask yourself: Who's asking? Maybe it's just a question of time before we run out of questions? Hmm, that's a good question. It's OK to ask, right? Just asking!

Our idioms, like our daily lives, are full of questions big and small; good and bad; easy, hard and frequently asked. It's the questions we ask, and those we fail to ask, that determine whether we gain the insight we need to do our jobs. But questions are about much more than developing or getting smarter.

A basic principle built into democratic structures is that citizens can question decisions made by their elected officials. And we, as individuals and societies, form our perception of what is right and wrong by means of question-based opinion polls, interviews and interrogation techniques.

Even so – or perhaps for this very reason – people rarely call questions into question.

DOING WHAT COMES NATURALLY

The way we start conversations by asking "May I ask you something?" is a good example of how naturally *asking* comes to us. It's not just the way we're always asking. It's the way we do so without noticing, and apparently without being able to stop. Asking questions is as natural as breathing.

At least that's the immediate impression one gets from reading the books and articles written about questions. The German philosopher Hans-Georg Gadamer is among the few who have analysed the essence of the question, and he holds that the structure of the question is presupposed in all experience. What he means is: Even when we're not asking questions, we're still relating to ourselves and our surroundings in a questioning manner.

Consider a preverbal child who crawls over to a ball, picks it up with her hands, licks it and turns it this way and that. Gadamer would say she is asking the question, "What's this?" When the same child throws the ball down and follows it with her eyes, she is exploring the question, "What can a ball do?" In this way all human actions can be understood as acts of questioning, and humans can be regarded as 'question animals'.

But if *Homo sapiens* is the questioning animal, how do humans differ from other animals? When a curious dog sniffs your handbag, is it not questioning, just like the little child? And what about the horse pressing against its owner to reach the carrot she holds in her hand? Is it not asking "Wasn't that for me?"

Experts on questions would reply, "No." Animals differ from human beings precisely by *not* asking questions. It's humans who interpret such animal behaviour as questioning, and we do so precisely because the question is an essential part of our own being – not essential like breathing is to all living creatures, but essential as a way of being in the world. Asking is a way of *Being*: the human way of being, in philosophical terms. It should not be understood as an action that can be performed more or less explicitly by more or less conscious beings.

The question is the essential characteristic that distinguishes human beings from animals – and, for that matter, from artificial intelligence and gods. The German philosopher Martin Heidegger explains this point by saying that human beings are the only beings who call their own being into question. Humans consider the possibility that they could be different – or cease to *be* at all – and they do so precisely by asking. "Who am I?" "Why am I this particular something and not something else?" "What does it mean to be – and not to be?"

Since none of these questions can be answered by anyone but ourselves, each and every one of us has to ask them. We have no choice. Thus, asking questions is not merely what distinguishes us from animals, artificial intelligence and gods. Questions also define us as human beings.

Questioning is a basic part of the human condition, and it applies to all people at all times. Questioning is that which cannot be otherwise, and which is therefore constant. Questioning, according to Heidegger, is the one thing we cannot call into question.

QUESTION — QUEST — TRUTH

Heidegger was by no means the first to give *the question* a central role in his understanding of what it means to be human. Nor was he the last. In Western thought and civilisation, it is an age-old assumption that asking questions is inextricably linked to being and developing as a human being.

This is why questions are the staple ingredient in all teaching, coaching and therapy. It is also why scientists across disciplines agree that questions are the key to becoming smarter. Although only a handful of researchers speak and write about questions themselves, they will unanimously confirm – if you ask them directly – that it is impossible to develop new thoughts and ideas without asking questions. Questions are the stuff thinking and development are made of; and it's thanks to questions that we have access to 'the good life' and to 'true knowledge'.

That, at least, is the assumption philosophers have gone by ever since Plato laid the groundwork for Western thought in ancient Greece. Most of Plato's works are constructed as dialogues between Socrates and various interlocutors. They are fictional transcripts of conversations where Socrates uses leading questions and answers to guide his followers to insight.

According to Plato, the things we experience with our senses are *phenomena*, and he understands phenomena in the light of *ideas*. For Plato, a horse is not a horse because it has four legs and a tail – as so many animals have – but because it takes part in the idea of the horse. Like all other ideas, the idea of the horse is part of *the world of ideas*, which differs from *the world of phenomena* by being inaccessible to sensory perception. Instead, we can come to cognitively know, realise and recognise the world of ideas, and we do so by means of questions. In the dialogue *Phaedo*, Plato has one of the characters summarise Socrates' point as follows:

If people are asked questions and someone questions them properly, the people, of themselves, describe everything as it is.

The idea of the human being is to know the good, the beautiful and the true, yet no human being is pure idea. Like the horse, we are also phenomena that can be seen, tasted, smelled, heard and touched. We must accordingly rely on questions in order to recollect the world of ideas from which we, and all other phenomena, derive our reality.

In other words, Plato designs all his dialogues around Socrates' questions not because he is interested in questions as such, but because he is interested in the answers that may surface by asking the right way.

For Plato, it is the answer – understood as our ability to recollect the good, the beautiful and the true – that is essential. Although we cannot recollect without seeking, and although we cannot seek without questioning, questions are always merely a means to achieve an end. A means that disappears as soon as the goal (here: knowledge or realisation) has been achieved. This has led some to think that Plato regards questions as the opposite of knowledge: as a human phenomenon that is to be transcended.

So we have Heidegger on one side, Plato on the other. Heidegger regards questioning as an essential characteristic of being human, one we cannot call into question because we are 'always already' questioning. Plato regards questions as a human phenomenon we ought not to spend our time asking about, inasmuch as, like all other phenomena, it draws attention from what truly matters: the idea of the good human being.

Heidegger and Plato thus agree that what is essential to the question is something other than the question itself. It is, instead, the human being – understood as a *mode of being* or an *idea*, respectively. Despite their very different approaches to questions, and use of them, Plato and Heidegger agree that the question of the question must be understood in the light of what it means to be human.

THE PARADOX OF THE QUESTION

Plato and Heidegger represent two different currents in understanding and using questions within Western philosophy and the history of ideas. Without asking themselves the actual question, philosophers have provided two main answers to why anyone rarely calls questions into question. First, it is because we *cannot* do so: It's always already too late to avoid asking. Second, it is because we *should not* do so: There's always already something else more important to ask about.

Both of these answers presuppose, however, that the question of the question dissolves within the question of the human being's essential way of being and developing. It is precisely the disappearance of the question about the question, and the fact that it's the question that dissolves itself, that set the scene for the French philosopher Jacques Derrida.

Unlike Plato, Heidegger, Gadamer and others who dealt with questions more or less directly, Derrida actually asks why we do not call questions into question. He also provides an answer – indeed, one that differs from Plato's and Heidegger's by being grounded not in history, but in the question itself.

According to Derrida, the reason neither philosophers nor scientists nor anyone else calls questions into question is that even though it sets something out in the open, a question always keeps something else hidden.

Now this is where things get technical. Not in the sense 'difficult to understand', but in the sense 'technological'. You see, there is a technology associated with asking questions.

The technology of the question lies in the fact that it forces the issue at hand to reveal itself in a certain manner. For example, when we ask what a horse is, we presuppose that a horse is a *something* that can be understood and de-scribed. If we instead ask how a horse differs from a human being, we presuppose that the *something* that a horse is must be understood in relation to a *something else*.

Whereas, in the first case, the horse reveals itself as something-in-itself – that is, as an absolute entity – in the second case, it is related to something else and thus reveals itself as a relative entity. In other words, the horse appears not only differently, but essentially differently, depending on how we ask about it. Put differently: The questions we ask, and don't ask, define our understanding of and relationships with our surroundings.

So what is concealed when we ask a question, and hence not called into question? The question's presuppositions. When we ask as we do, we are not aware that *something* will reveal itself in a certain manner, so we are equally unaware that we could be asking differently – and that *something* could thereby reveal itself differently. In short, we are not aware of what the question does.

Posing a question means positioning oneself in a certain way in relation to the object of the query. One poses; posits; takes a position. With this positioning, one is always in the process of answering. This is why I was able to deduce earlier – in the case of the question, "May I ask you something?" – that it resolves the issue at hand before raising it.

This is the paradox of the question: Even as it opens up new ways of seeing and understanding the world, it simultaneously establishes a framework for, and thereby a limit to, understanding. If different people and cultures turn out to have different ways of relating to and using questions, then the line between what is comprehensible and what is incomprehensible will necessarily be drawn in different places, and the seeds of misunderstanding will have been sown.

Questions are just as revealing as dreams, or even more so, the German-American phenomenologist Erwin W. Straus once remarked. But if, in asking, we are always answering, then what we are revealing is always ourselves. Does this mean we cannot, should not, or must not call questions into question?

No, it means, instead, that the question is the key to understanding why we think and act as we do. Questioning is not a human characteristic that always unfolds in the same way, and so cannot be otherwise. Rather, questions are what reveal – precisely because they could be otherwise – how we perceive ourselves and our surroundings. In other words, questioning is not just what distinguishes us from animals, artificial intelligence and gods, but also what distinguishes us from each other.

There are many ways to ask questions, and thereby to think and be. The question is: What does it reveal about Western thought and civilisation that we relate to and use questions as we do? What does it mean for our relationship with ourselves and our surroundings that we have historically understood the question of the question in light of the question of Man? And what opportunities do we have for exploring whether there are other options?

THE
FIRST
QUESTION

AN ANCIENT INVENTION

It's always a good idea to start at the beginning, so if we wish to understand how the way we ask questions has impacted Western thinking and civilisation, then we must ask: Where does our understanding of questions come from? Who asked the first question in the history of humankind – and how does it affect us today?

Like so many other questions about questions, these have no unambiguous answers. We cannot date or otherwise document when, how, or why we humans began to ask questions. However, we can state that questions are an ancient invention.

Questions have probably been around as long as language has. Depending on how we define 'language' and 'questions', it may also be the other way around, with language arising as a consequence of humankind's inquisitive nature.

It is well known that language fails, in many respects, to give adequate expression to the things that affect and inspire human beings. Expressions such as 'speech is silver, but silence is golden' surely did not arise out of thin air.

Rather, basic human phenomena such as surprise and doubt, ignorance and curiosity – which all articulate a questioning mode of dealing with the world – may well be the reason why human beings developed language.

Whichever came first, language or the question, clearly we must look into our distant past to pinpoint the first question in the history of humanity. Perhaps we must even go back to Adam and Eve?

WE ARE ALL SINNERS

Down through the ages, myths have helped people find meaning and direction in what cannot be documented. Mythology – more specifically, the myth of the Fall as described in the Bible – can shed light on our understanding of questions and how we use and relate to them.

According to the Danish philosopher Søren Kierkegaard, the myth of the Fall reveals no more about Adam and Eve than it does about all other human beings. "Much as when Adam lost innocence by becoming guilty, so too does every human being lose it," Kierkegaard writes in *The Concept of Anxiety* from 1844. This means that we who are alive today are neither spectators to sin nor innocent heirs to it. Rather, we participate in it actively and contribute to it. Put another way, we are inherently responsible.

Here, as in many other cases, language offers us a gift. Responsibility means 'the duty to respond'. Fittingly, in effect the English word 'responsibility' contains the word 'response', just like similar words in Italian, French and other languages, as exemplified by the word pairs *ansvar/svar* in

Danish, and *Verantwortung/Antwort* in German. In many Western European languages there is a connection between responsibility and responding, and if human beings have a duty to answer, then there must also be someone who asks the question.

QUESTIONS TRIGGER THE FALL

Who could this someone be? The French philosopher Emmanuel Levinas has an idea. According to Levinas, the first question in human history was not a question posed to God and about God by Man, but rather a question posed to Man and about Man by God. This, however, is not entirely consistent with the myth of the Fall.

According to the Bible, God created the heavens and the earth, light and darkness, day and night – and not least 'Adam', 'the man', meaning humankind. It is God who forms this human being from the dust of the ground, God who breathes the spirit of life into his nostrils so that he becomes a living being, and God who commands Adam not to eat of the tree of the knowledge of good and evil. Then something happens. Suddenly, after having passively been formed, having life breathed into him, and being commanded not to eat of the tree of knowledge, Adam gains an active role in the creation story.

It is not good, declares God, for the man to be alone, so to end his loneliness he receives not only animals and birds, but also language. Adam is made responsible for naming the various creatures, and it is while naming the various animals that Adam becomes aware – and makes God aware

– that something is missing: He has no companion. God quickly supplies one: Eve.

The arrival of Eve not only brings Adam a new and important relationship, but also weakens his relationship with God. Previously, Adam had awaited God's commands and followed them blindly. Now he stands with Eve: "Therefore shall a man leave his father and his mother, and shall cleave unto his wife" (Genesis 2:24). Adam, and every man after him, leaves his maker(s) for the woman with whom he will form "one flesh".

Here it is already relevant to speak of loss and a feeling of betrayal – as no doubt many fathers and, not least, mothers(-in-law) will attest. It isn't pleasant to be left by a person whom you've created and loved, and whose needs you've done everything to fulfil. No wonder some parents feel the urge to shout, like God: "What is this that thou hast done?" at a woman for whose sake the man they created is now defying their commands.

But what triggers the Fall is neither loneliness nor language nor the feeling of loss, nor even the woman for whom Adam forsakes God. Only when the sly serpent approaches Eve do things go wrong. Unlike God, this snake does not command Eve to behave in a particular way, nor, as in Adam's case, to give names to the birds of the air and the beasts of the field. Instead, the serpent approaches Eve by asking a question. "Did God really tell you people that you couldn't eat from 'any tree' in the garden?"

QUESTIONS ENTAIL RESPONS(IBILITI)ES

At this point in the story, when the serpent addresses Eve, neither God nor Adam has yet put their language and speaking ability to use in asking questions. Instead, it's the serpent who asks the first question in human history. It's the snake who opens up the possibility of reflecting on the commandment that God has given; the snake who calls the omnipotence of the Lord God into question.

The serpent's question tempts Eve with the possibility of learning what good and evil are. It is this possibility that makes the difference. The possibility of answering for *oneself*: not for God and not for Adam, but for *me, myself*. Here, *I*, Eve, must take a stand on God's commandment. And *I*, Eve, choose to defy it.

The nature of the question dictates not only that Eve *can* respond to it, but that she *must* do so, and with her response comes responsibility. Not the absolute, omniscient and all-encompassing responsibility God has, but the relatively limited and self-absorbed responsibility the serpent evokes. In this there are several consequences that have philosophical significance.

The first is that questions set the framework for the responses that may be given to them, and hence also for the responsibility that can be taken as a consequence. Once the serpent has asked his question, Eve can no longer *not* desire to know good and evil. This leads to the next philosophical consequence: No matter how hard we try, as human beings, to understand and strive for the good, our contribution will always be relative, limited and defined in terms of ourselves.

In this way, the discussion of human beings' conditions and constraints is linked to asking and being asked questions. According to the myth of the Fall, all human misery was triggered by a question. It is a question that leads Eve to believe she can rise above her own constraints and desires, and understand more than she really can. Eve is not God, but the question leads her to believe she can be as good and just as God. The question ignites courage in Eve, or rather the foolhardy idea that she is more than just a human being.

It is quite thought-provoking, this connection between being asked a question and being confronted with one's own constraints as a human being, and seeking to reach beyond them. So is the fact that the source of the question is the serpent – or, perhaps more to the point, that the question does not come directly from God.

When humanity is confronted with its first question, God is out of sight and out of reach. The mere fact that the question emerges in God's absence may itself be significant: Only because God is otherwise engaged does it become possible to call into question his command not to eat of the tree of knowledge.

We don't know where God is or what he is doing as the serpent approaches Eve. All we know is that God has nothing whatsoever to do with the Bible's very first question, which occurs during a tryst between the serpent and Eve: the animal who – unlike the other animals – has language, and the woman who – unlike the man – is created to socialise.

EVERYTHING UNDER THE SUN

To Eve, nature manifests itself in the form of the serpent as something that can communicate with her, and persuade her. It is precisely in nature's persuasion of humanity – a persuasion that leads humanity to feel superior to nature, and capable of issuing proofs to it – that the danger lurks. The danger that we will become blind to our own limitations, and so confuse self-love with divine omnibenevolence.

Nevertheless, God does not break in and dispel the misery. God does not merely *permit* humanity to commit the worst mistake of all time; he seemingly *invites* it. Why? Perhaps because the whole point of the Christian human being is that it bears the transition from innocence to guilt within itself, and with it, the transition from ignorance to knowledge? According to Kierkegaard, sin is not an expression of any imperfection in us, but of an originality that is good by virtue of being evil. The relation between good and evil must be understood in the same way as that between the new and the old: They are conditioned by and grow out of one another.

This inner connection between opposites is a consequence of God creating the world in pairs: heaven and earth, light and darkness, day and night, man and woman. God's creation is dialectical. This means that even the most distant phenomena 'commune' with each other.

Everything under the Sun is connected in one great, mutual dialogue driven by questions and answers.

From this perspective it is incorrect to suggest, as I did above, that God has nothing whatsoever to do with the first question in the Bible. God is present there precisely because he is absent. We could also say God designed humankind so he doesn't even need to be present for us.

In this reasoning, when Eve engages in dialogue with nature she does not defy God, but completes his dialectical creation. She demonstrates the necessity of moving between questions and answers, thereby completing the transition from innocent to guilty. Or perhaps just: *the transition*.

According to the French philosopher Paul Ricoeur, it is highly significant that the Bible consists of both the Old and the New Testament. The point is, not until an event becomes temporal – that is, human – does it become meaningful.

It is not enough that God is omnipresent, found every*where*. He must also be in every*thing*. This requires a transition: from the Old to the New Testament, from the almighty Father in heaven to the powerless Son on the cross. God is in Jesus, in you, and in me. He is also in Eve when she responds to – takes responsibility for – the question of good and evil. That is why this story takes place on the very first pages of the Bible: because the point in time when humans take responsibility for their own role in, and significance for, Creation is the prerequisite for everything else.

So can we use our insight into the myth of the Fall to learn about other cultures? Can we better understand how other human beings relate to themselves and their surroundings by investigating the role of questions in the essential religious writings and myths undergirding their cultures? Can the first question in the Quran, for example, tell us something about how Muslims understand humankind's role and responsibilities?

BLIND SPOTS IN WESTERN THINKING

I'm the one asking all the questions in this book. Therefore, my personality and my way of thinking are also revealed in them. What do I reveal by asking about the first question in the Quran? I reveal my own (Western) presumptions in asking that question as I do.

My first presumption is that it's a good idea to start at the beginning. That is: I think chronologically. My next presumption is that there is a cause for everything. I expect not only to find links between different relationships – for example, between the role of the question in the Quran and Muslims' understanding of the role and responsibilities of humanity. I also expect that such links can be understood and explained. Finally, my third presumption is that the responsibility for understanding and explaining such links lies with me.

Hence, even though I will not learn much about Muslims by asking my question about the first question in the Quran, there is a purpose to asking it. This purpose is to direct my attention to the limitations of my own culture.

The cultural presumptions that lead us to ask questions as we do are so natural to our Western way of thinking and being that we forget to question them as well. But our cultural presumptions are by no means universal. I learned in a preface to the Quran (translated from the original Arabic) that the Quran is not arranged chronologically.

Without chronology, there is no beginning. This makes it difficult to determine who posed the first question in human history according to the Quran, and what that implies. Or perhaps it is not so difficult? Although the Quran is not chronologically arranged, there is only one entity that could have asked the first question in the history of humanity. The one who asks all questions and supplies all answers: God.

A QUESTION OF CULTURE

The words of the Quran appear as the word of God. They are recited in Arabic from the heavenly scripture that, for Muslims, contains the wisdom of the world, and which was granted to Muhammad in a series of revelations. This means that the Quran is characterised by orality.

The reason I can unequivocally answer my question about who asks the first question in the Quran is simply that there is only one speaker in the text: God.

This is also reflected in the passage that describes the naming of the various beings God has created. The Bible says that God leads the beasts of the field and the birds of the air to Adam to see what he will call them: "whatsoever Adam called every living creature, that was the name"

(Genesis 2:20). The Quran says that God taught Adam all the names.

As a Westerner, I have been trained culturally to believe that texts should be interpreted and explained. Here it is tempting to begin analysing the Quran, but according to Muslims, I ought to be careful.

I do not read Arabic – the language of the Quran – and worse, I am a human being. This means my interpretation will always be incomplete and incorrect. Put another way: The message that the Quran is the word of God must be taken literally.

In the Quran, God appears variously as a first-, second- and third-person voice, but Muslims believe it is always God himself who speaks through his prophet Muhammad. This means that dialogue – understood as the dialectical interplay between questions and answers – has a different meaning in the Quran than it has in the Bible.

According to Mustansir Mir, a professor of Islamic studies, there are five types of dialogue in the Quran. The first type is dialogue between a prophet and the nation to whom he has been sent. The second type is dialogue between God and the prophets. The third is set in the afterlife, between the people of Heaven and the people of Hell, for example. The fourth is between parties consulting with one another about an important cause. Finally, the fifth type of dialogue is characterised by the fact that only one voice is transcribed, as when God addresses Satan. So if the Quran is the word of God, it follows that not only the fifth type but all five types of dialogues are transcripts.

God's transcripts. No matter whether a particular utterance is attributed to a prophet, to the infidels, or to God himself, it is God's utterance all the same. Put another way: Because the Quran manifests itself as the word of God, it bears the mark of orality in the form of a single voice, a single perspective, a single answer.

This is unlike the Bible, which consists of a collection of writings by many authors, and so includes many perspectives. The Bible is not the word of God, but numerous prophets' and evangelists' interpretations of the word of God, which it is our responsibility to interpret.

What this difference reveals about Muslims' understanding of the role and responsibilities of humanity I cannot say, but it does call into question our Western way of regarding co-determination, dialogue, and human responsibility as self-evident. More than anything else, our relationship with and use of questions are expressions of our culture. A culture that places us and our responsibilities as humans at the centre of the world.

MAN'S INNATE POTENTIAL

This opens up another angle of attack regarding the first-question issue. Instead of looking for answers in myths and religious writings, we can ask about the role of questions in the individual human being's linguistic psychological development. That's exactly what a number of psychologists and language experts did in the nineteenth and twentieth centuries.

In 1900, the German psychologist Wilhelm Wundt

wrote that questions are a type of sentence that occurs in all languages. Because he considered language part of Man's innate potential, he regarded questions, too, simply as part of humanity's "natural equipment." However, the Austrian philosopher of language Friedrich Kainz disagreed. Unlike Wundt, Kainz regarded questioning as one of the highest and most complex functions of language.

Kainz cited studies in developmental psychology as being strongly suggestive that the human function of questioning is the last to be developed, and also the first to be abandoned when an individual or a people comes under pressure. To illustrate how complex and multifunctional an activity asking questions is, Kainz constructed a hierarchy of human expressions. This hierarchy applies to the linguistic psychological development of both individuals and humanity, and it consists of four stages of development:

In the first stage, one is unable to express anything but one's presence, as with a screaming toddler. In the second stage, one can appeal to and influence others, as God does when issuing the ban on eating from the tree of knowledge.

In the third stage, one can communicate information that refers to a specific object or matter. Adam embodies this stage when naming the living creatures God has created. In the fourth and highest stage of development, one is able to draw a connection between the speaker, the listener and the object or matter spoken about. That is what the serpent does when posing its question about God's prohibition to Eve.

Kainz points out that the lower functions of language can develop in animals and in toddlers, who are still underdeveloped. The little child screams, and the dog barks, to draw attention to itself and to its needs. Not so with questions, which require a high level of development.

One of Kainz's arguments is that questions presuppose, as a minimum, an understanding of the difference between truth and falsity, or between reality and fiction. This is a clear extension of an earlier assumption in the psychology of language: that yes/no questions were developed before 'supplementary questions' – a fancy name for questions starting with who, what, when, where, how and why.

This approach to and understanding of questions may seem outdated, or even mechanical, but Kainz's hierarchy of utterances, and his considerations about the relation between the question and other forms of expression, are by no means insignificant. They also reveal things about our Western thinking and civilisation; about the way we have understood and used questions historically; and about the way we understand and use questions today.

THE HISTORY OF THE QUESTION

QUESTIONS WITHOUT ANSWERS

141. That's how many questions my relatively ordinary family asked one ordinary morning. From 6:50 to 7:40, two children and two adults averaged a little over 35 questions each.

Suppose each of us asks about 40 questions per hour for about eight hours a day (reasonably assuming there are intervals at school, work meetings, football practice and the like with a significantly lower question rate). This yields 320 questions per family member per day – excluding all the questions we ask ourselves without verbalising them: "How long can I put off getting up?" "Do I really have to go for a run *today*?"

Not all the questions we ask ourselves and each other are of the information-seeking kind. Far from it, and one can certainly question how much knowledge and development they reflect.

"Are there any eyes in there?" my husband will ask our children while waking them up for school. "What's bugging you, girl?" I ask my dog as it gnaws at an itchy spot. "Why can't I stay asleep when it's still the middle of the night?" our seven-year-old daughter asks, meaning she finds it too early to get up. "How can a duvet feel so comfy?" our ten-year-old son asks as he stretches in bed, hoping to get two more minutes in his warm nest.

None of these questions live up to Gadamer's definition of a true question, for they demand no answer. The idea is not that our children should answer 'yes' or 'no' to whether they have eyes beneath their tightly shut lids, and it would be odd indeed if our dog looked up at me and growled, "Yes, these fleas are really annoying." Nor does our daughter expect an answer to her question. Believe me, if she did she would say so. And our son would certainly protest if I began a disquisition on the various meanings of 'comfort'.

So why do we ask such questions? What is the purpose of asking questions we neither expect nor want to have answered?

THE THREE BIG E'S

Questions obviously have multiple features and aims. While the questions I just mentioned serve as modes of expressing oneself and one's needs, they also serve as social markers of attention and care. Consider these questions: "Do you have gym class today?" or "Are you buying the gifts?" or "Do you think we have enough leftovers for supper?" or "Will you be eating at home tonight?"

When we ask questions of this type, we hope for a response that will clarify certain practicalities for ourselves and for our respondents. That's why failure to respond immediately to such questions can make you very unpopular – because your answer is needed to determine how the questioner should act.

That's not the case with such questions as "What are you doing?" or "Is this a good place to sleep?" or "Why is it so dark here?" These reflect situations where the questioner is considerably less oriented toward, or dependent upon, an answer. And it is certainly not the case with questions of the type: "Can you find me a scrunchie?" or "Who's going to be the first one out the door this morning?" or "How about a hug?" Here, what matters for the questioner isn't the answer to the question, but the *action* it prompts.

Of the 141 questions posed on an ordinary morning in my ordinary family, only a single one was aimed directly at increasing the questioner's knowledge. As it happens, that question was never answered. Gazing at the ball-point pen I was holding in my hand, my son asked: "How can the colour keep coming out of the tip?" That could have been an interesting question to explore together, but we didn't, and we rarely do.

Nevertheless, many philosophers regard this as just the sort of question that distinguishes human beings from other living creatures: questions seeking an answer that explains how the world works.

Straus, the phenomenologist mentioned earlier, believed that questions can be divided into three groups, depending on whether the questioner is oriented toward himself, an object or another human being. This understanding of and approach to questions corresponds to the three main topics in Western philosophy, which can conveniently be nicknamed the "three E's": existence, epistemology (the theory of knowledge), and ethics.

Existential philosophy is concerned with what is revealed about the questioner by his orientation toward himself. *Epistemology* is concerned with what is required before the questioner can be said to have attained true knowledge of the object or matter toward which he is oriented. *Ethics* is concerned with the responsibility that the questioner assumes when he orients himself toward another person.

Hence, no matter what we are asking about, and no matter when, why, or how we ask our questions, we are always orienting ourselves toward something. The question is what our questions reveal. Do they reveal the 'something' we are oriented toward? Or do they reveal the very fact of our orientation?

If the former is true, then philosophy is all about asking the right questions in the right way in order to gain the right understanding of the nature of the world. If the latter is true, then philosophy is all about exploring and challenging the conditions for our asking and orienting ourselves as we do.

The notion that our questions reveal a 'something' independent of both questioner and respondent has undoubtedly played the dominant role in the history of the question – a history that has not yet been recorded, although a few initial attempts have been made.

THE HISTORY OF THE QUESTION

The Dutch philosopher C.E.M. Struyker Boudier has

explored the various traces left by the question since the foundations of Western thought were laid approximately 2,500 years ago. There are many such traces, pointing in a variety of directions across a variety of historical periods; but the dialectical movement between opposites, which we also explored in connection with the myth of the Fall, has always played a role in our understanding of questions.

As a method, dialectics works by setting claims and counterclaims in opposition to one another, and then using questions and answers to reveal that the two opposed claims in fact cohere. Ultimately – according to the dialecticians – the contradictions dissolve, and one achieves knowledge.

This is how Socrates proceeded, and how Plato believed all philosophers should proceed. According to Plato, dialectics – the art of conversing and arguing – is distinguished by its focus on attaining knowledge of *the ideas*. Unlike rhetoric, dialectics is neither a game with words nor a method for impressing or persuading one's listeners. According to Plato, the dialectical method is the only true form of cognition.

These days, few agree with Plato's conception of knowledge as the recollection of inborn ideas. However, his notion that the road from ignorance to knowledge travels through questions, and that we can achieve true knowledge only by continuously asking sharp and relevant questions, has in fact survived.

Ever since Aristotle – who was a student of Plato – and up to our day, philosophers have incessantly repeated that if knowledge is not an answer to a question, it is not true knowledge. Thinking has been regarded as a dialectical exchange of questions and answers. It's all about knowing and not knowing, and about asking for the knowledge we do not yet have. From this perspective Gadamer, commenting on the Socratic-Platonic dialogue, notes that the art of true and good questioning is more difficult than the art of answering.

This view also prevailed during the period from 1100 to 1500, when the so-called scholastic method dominated philosophical self-understanding and debate. 'Scholastic' literally means 'belonging to the school', and the approach covers a wide range of ways in which the scholars of the time explored, challenged and, in particular, settled debates or 'cases' by means of questions. This took place not only behind closed doors, but also, and especially, at public events with audiences from all walks of life.

I imagine a time when, rather than being entertained by amateur talent shows or videos on social media, people would flock to be enlightened by scholars like my colleagues. A time when, rather than enthusing about a song and dance, the spectators applauded logical rigor and precision. A time when, rather than vying for likes and followers, contestants battled over – and attained – insight into everything from the fundamental principles of physics to the nature of the soul.

Such concepts as "battle" and "attainment" are wholly appropriate to use in connection with scholasticism. During this period, verbal jousting, dialectically opposing claims with counter-claims, was a worthy pursuit. Imagine a contest inspired by Aristotle's theory of the soul: One participant would argue for the immortality of the soul, while the other would contend that the soul is inseparable from the body. The most convincing and eloquent of the two would be declared the winner.

Sometimes such bouts were almost like a tournament where participants would defend their cause, and their personal reputation, by means of critical questions and answers. Other bouts proceeded almost like a trial where participants were tasked with solving a problem or riddle with the aid of logical arguments.

Under these conditions, the participants appeared as lawyers, each arguing for his preferred solution with the aim of securing a 'judgment'. This was issued, as in our legal system today, based on what scholasticism called a *quaestio*: a questioning exploration that in the dialectical tradition, as in American films, is known for uncovering "the truth, the whole truth, and nothing but the truth".

NEW QUESTIONS, NEW METHODS

Aristotle was not merely a great source of inspiration for the philosophy of the late Middle Ages. He is also, to no small extent, responsible for the contempt for questions that has persisted throughout much of history. What Aristotle did was to classify questions as expressions whose truth or falsity cannot be established.

Historically, this has meant that the question as an object of study was confined to linguistic research, while large swaths of the philosophical tradition focused almost exclusively on expressions bearing truth-claims. So it is that even though Plato considered it the most important factor in achieving true knowledge, the question became a minor, all but neglected topic in philosophy.

This was especially true in the 1600s, with the advent of the French enlightenment thinker René Descartes, who wished to expel dialectics from philosophy because it does not lead to reasoned knowledge. He had absolutely no respect for curiosity, which he held leads to a disorderly search for nothing at all. Instead, one ought to search systematically for something specific. Thus, the emergence of modern science and philosophy is not only linked inextricably to the emergence of new questions, but also, and especially, to the emergence of new *ways* of asking questions.

Descartes mocked the kind of war or battle that is built into the scholastic method. He argued instead for a strictly logical approach that includes what he regarded as a *correct* way to ask questions. The subject of one's investigations had to be arranged in the proper manner so that obscure and complicated questions would be broken down into simple, straightforward sub-questions, all of which could be answered.

Many other famous philosophers have also contributed to the general perception that what is interesting about questions is not the questions per se, but the truth (or lack of truth) in the answers one can reach.

Whether they have regarded dialectics as the only true

form of cognition, like Plato, or have expelled dialectics from philosophy, like Descartes, most Western philosophers have engaged with the question in the light of the answer. That is to say, they have engaged with the 'something' about which the question seeks to gain true knowledge. In this way, the history of the question has become the history of answers. Or put another way: Answers have diverted the attention from the question of the question.

THE QUESTION AND HUMAN FREEDOM

In the nineteenth and twentieth centuries, after focusing solely on the importance of the question of knowledge and cognition – the epistemological aspect of our three big E's – philosophers began to take an interest in the question's role for the other two E's in philosophy: existence and ethics. In France the existentialist Jean-Paul Sartre and the ethicist Emmanuel Levinas (and others, elsewhere) saw the question of the question as a question of human freedom.

But freedom is many things. For Sartre, freedom means fumbling hands and searching gazes. It is in searching, fumbling, hesitating and making attempts that we relate freely to what is. The fundamental freedom of humankind can be seen, for instance, in the fact that even as we are considering our own hunger, we are also considering the possibility that we could be full, as we search for something edible. According to Sartre, we do this precisely by asking with the body.

Not so for Levinas. He is interested not in humanity's fundamental freedom, but in the moral consciousness that is possible to achieve when someone calls my freedom into question. For Levinas, moral consciousness is the highest possible form of consciousness, achieved only in the encounter with another human being. With his questions, the other human being holds me *responsible* – as discussed earlier – and this responsibility defines me as a human being.

Put bluntly, while Sartre regards the question as an expression of human freedom, Levinas regards human freedom as something that needs to be questioned. For Sartre, each of us *is* a human being. For Levinas, we *become* human beings together with others.

However, both these thinkers are far removed from the view that philosophy is about asking the right questions in the right manner, thereby achieving the right understanding of 'something'. Both take their cue from the second possibility: that what our questions reveal is that we are orienting ourselves in the world.

QUESTIONS
AND BEING
HUMAN

A QUESTION OF TIMING

Struyker Boudier describes the historical development from Plato to Descartes as a time when old questions about 'why' and 'for what purpose' were replaced by new questions about 'how' and 'what'. But if we want to become more aware of why we ask and orient ourselves as we do, what questions should we ask? And does it even make sense to go on regarding some questions as better or more right than others?

Different questions certainly set different frameworks for how they can and cannot be answered. For example, the only way to answer the question "Who's going to the party on Saturday?" is to list the names of those expected to attend. The question "Why is there always so much booze at your parties?" can be answered with anything from "Because my dad owns a wine and spirits shop" to "Because girls are more fun when they're drunk".

The two questions above are hugely different. The first questioner clearly has prior knowledge of the issue at hand, while the second questioner may well be completely ignorant as to whether an answer is appropriate or not.

Put another way, while the first question is determined,

and thereby limited, by a concrete context – a particular list of invitees for a particular party on a particular Saturday – the second question is not determined, and thereby limited, by anything at all. It is, or seems to be, entirely up to the respondent to decide what explanation he will give for the abundant amounts of alcohol.

We can thus distinguish between two categories of questions. On the one hand, there are contextually de-termined 'who', 'where' and 'when' questions. Like yes/no questions, these are typified by the questioner being familiar with the framing of possible answers. On the other hand, there are explanation-seeking 'why' and 'how' along with certain 'what' questions, which are independent of the questioner's own knowledge of the answer.

This links some questions to our ability to draw upon our experiences, and links others to our ability to think rationally. Briefly put, we link different question words to different forms of cognition.

WHO — WHAT — WHERE — WHY?

Heidegger founded his whole thinking on the claim that Western philosophy has asked the wrong questions, thereby taking a wrong approach to what it means to think and be. According to Heidegger, the most important thing noted about a question is not what is being asked, but who is ask-ing, and why. Heidegger thereby moves between and across the two categories of questions. That is no coincidence.

For Heidegger, the interesting aspect of *the questioning existence that human beings are* is not its subjectivity, but its being, understood as being something and not something else. This is true of humans, but it also applies to everything else. Accordingly, Heidegger holds that the way to 'something' necessarily goes through 'someone'. Put differently: We cannot recognise what something is or how it works until we have experienced that neither we nor anything else is just anybody. Rather, like all other phenomena, we are first and foremost our particular selves.

Heidegger thus implies that all questions are contextually determined – inasmuch as they are necessarily asked by someone – which is the point he criticises Plato and his successors for missing. By asking how the world works and so orienting themselves toward 'something', Western philosophers have forgotten that the precondition for cognition is recognition.

Just because a little child gets it right once, correctly calling the fruit in its hand an 'apple', that hardly means it has come to know what an apple is. Many an eager parent has had to admit that after passing their child a banana, they got exactly the same reaction: "Apple!"

So long as the child keeps calling apples by other names, and calling all sorts of other things 'apples', then she simply hasn't yet come to know what an apple is. Only when the child has repeatedly recognised an apple as an apple, and consistently experiences it as 'a particular something', as opposed to 'something else'; only then can she be said to have acquired knowledge.

We can also say that recognising 'something else' presupposes an experience of 'the same as'. Or rather: Only when we begin to relate to 'ourselves' do our surroundings begin to take shape as 'themselves'. This does not happen at a certain point in time, but over time, which is why cognition is conditioned and limited by experience.

MADE OF FLESH AND BLOOD

In Plato's dialogue *Republic*, Socrates speaks of the importance of teaching children to ask and answer questions. He is not speaking about real children of flesh and blood, however, but about the children his conversation partner is training and educating 'in thought'. Here, Socrates is speaking about the ideal upbringing of the ideal child, and about the question's role in realising this ideal.

The question Socrates asks his interlocutor is: If someday you had a chance to instruct these children in the real world, wouldn't you teach them to ask and answer questions as insightfully as possible? His interlocutor promptly replies: Yes!

Of course children must be encouraged and stimulated to ask insightful questions. I'm with Socrates on that. Still, my son never got a response to his explanation-seeking question about how coloured ink can keep coming out of a pen. On that ordinary morning, an ordinary event took place: I didn't even let him know I'd heard his question. Why? Because, unlike Socrates, I'm not training and educating in thought, but in practice.

I'm not ideal, and neither are my children. We're flesh and blood. That affects how I raise them, for it means that no matter how much I read and write about the signifi-

cance of the question in realising the ideal, I am subject
to certain conditions and limitations: constraints on my
available time, my wishes and my abilities, and the fact that
someday I will die. I am neither a pure idea nor a pure soul,
but a body bounded in space and time. As my children are
well aware, I cannot know, remember or answer everything. They know this is a condition we share. Sometimes
I even suspect that's why they keep asking me: in order to
share, and force me to share, the conditions and limitations
we have in common.

You see, there is a connection between our conditions
and limitations as human beings on the one hand, and
asking and receiving questions on the other. We can easily
forget this connection when cultivating the idea that
questions lead to truth, wisdom and ideal child-rearing.
Meanwhile, in practice we become acutely aware of the
connection, and of our own constraints, when confronted
with questions we cannot adequately answer.

WHY THIS? WHY THAT? WHY NOT?

For many Danes, the question of the question calls to mind
Spørge-Jørgen ('inquisitive Jørgen'), a popular nursery rhyme
and children's song by Kamma Laurents, appearing in
1944 as a book illustrated by the beloved cartoonist Robert
'Storm P' Petersen. Let's call the protagonist of this Danish classic 'Asking Andy', since *at spørge* means 'to ask' in
Danish and the boy's claim to fame is asking silly questions
from dawn till dusk about everything that crosses his path,
or his mind. Asking Andy's deluge of questions continues

until one day his father, fed up, spanks him and sends him off to bed. Here, the boy very sensibly wonders: "Why did Mummy take away my pancakes? // Why did Daddy ruin all the fun? // Why do spankings have to be so painful? // I swear my silly-question days are done!"

Some Danes deliberately don't read this book to their children, as they disapprove of the moral of the story. But what *is* the moral of Asking Andy? That children shouldn't be silly? Or that silly children should be sent to bed hungry? Or that adults can find it so hard to deal with a child's insatiable curiosity that they feel compelled to curtail it?

Personally, I can relate to the mental fatigue Asking Andy's parents feel at their son's perpetual, irrational, unanswerable question. "Oh, stop that nonsense!" As a mother, I must confess this answer has crossed my mind, along with: "Ask me something sensible! And for once, could you just *wait* for the answer, the *whole* answer, before firing more questions at me?" Luckily, I've never said these things out loud. If I had, my children would no doubt have asked me what questions are sensible, and why – and I would have no idea how to respond.

Certainly I can establish, as the nursery rhyme does, that it's more sensible for Andy to ask why a spanking hurts than why people milk cows and not chickens. But when I have to explain why the first question is more sensible, things get ugly. Isn't the only real difference that, being adults, we can answer the first question almost automatically, while the second calls for active reflection?

Does the sensible approach we adults prize so highly

– the link between cause and effect, between a spanking and a sore bottom – amount to anything more, anything other, than mere laziness? The French philosopher of science Alexandre Koyré would say it does not: Man is a lazy creature, a sloth, who only bothers to think when he cannot solve his problems by other means.

Adults do have the means to put a stop to Asking Andy's eternal questions. By sending him off to bed, they can expel him from a desirable social setting. They can also ignore him and act as though his questions are irrelevant to others. A very effective strategy.

Before long, Andy will start keeping his questions to himself. He will likely try to get another adult – a grandparent, perhaps, or a schoolteacher – to join him in exploring life and the world. If that doesn't work, within a few years Andy will forget the questions he once found so important. Instead, he will have learned how to answer questions just as adults do. He will have become … sensible.

Kierkegaard writes that he prefers to talk with those who are destined to become rational, sensible creatures, but cringes over those who have actually become such creatures. This is his reaction when seeing childhood questions replaced by adult answers. What is sensible is sensible because it cannot be otherwise, and so there is no reason to ask questions about it. Most parents have heard themselves say: "Well, that's just the way it is." But what makes children's questions so difficult for grownups to handle? Why is it so important for us to avoid active reflection when answering?

FEAR OF 'ASKING ANDY'

In this context active reflection, thinking, should be understood as exploring one's boundaries as a human being – which can be intensely challenging! One line in our old nursery rhyme has Asking Andy wondering why people don't have eyes in the back of their heads. What is he reminding us of here? That there are limits to what we can see, and hence limits to what we can comprehend.

As long as we only ask questions we can answer, never go beyond what we can handle, we can give ourselves and others the impression that we have existence, life and the world, under control. But Asking Andy's questions scare us. They remind us there are things we can neither control nor change. After all (in answer to another of Andy's odd queries), we didn't decide that nails belong on people's fingertips, not on their noses. That's just the way it is.

Asking Andy reminds us of that essential human feature: our constraints. We are limited in time and space, in perspective, in insight. We cannot freely choose our standpoint for regarding the world. As the German philosopher Friedrich Nietzsche puts it: "One hears only those questions for which one is able to find answers."

The question is: What determines which questions we hear, and don't hear? Is it a matter of training and education, as Socrates believed?

The American anthropologist Esther N. Goody saw many indications that this is the case. In 1978, Goody suggested a correlation between how early and how often questions are posed to small children, and how they develop and thrive as adults. She referred to studies showing that middle-class mothers ask their preverbal children more questions than working-class mothers do.

Goody, not wishing to jump to conclusions, awaited further studies, which regrettably were never done. Still, she strongly suggested that 'training questions' – "What's this?" / "Is it a dog?" / "What colour is the dog" / "Is the dog brown?" – play a decisive role in the child later developing the ability to ask such questions herself.

Training questions are manipulative. Their purpose, obviously, is not to prompt the child to respond, but to isolate vital contextual information. Goody believed it is precisely this ability to seek and isolate information that determines how the child will fare later in life.

She even went on to contemplate why pre-industrial peoples and societies were unable to develop. She reasoned, controversially, that it was because they hadn't learned to ask questions and were unable to seek out the information necessary to think and create along new lines.

Goody's studies are interesting because they show that different people handle and employ questions differently. But can the difference between developed and less developed societies really be explained through the former understanding and employing questions more effectively than the latter? And what does it say about Goody's results that she defines an effective question as one that seeks and isolates vital information?

Goody's premise, like that of Plato, Gadamer, and many others chiefly interested in questions in light of the answers that can be reached, presumes a direct connection between 'the good question' and 'the good life'. Goody thus bases her analysis of non-Western societies on the ideas and ideals of the West. The question is whether she had – and we have – any other options.

Inasmuch as a human is a finite being, the questions we (don't) ask, and those we (don't) hear, will always be conditioned and limited by our own location in the world. Perhaps that is the point: that instead of worrying so much about whether our questions are right or wrong, or sufficiently insightful, we should look more closely at the questions we actually pose.

It is noteworthy that only one of the 141 questions asked on an ordinary morning by my ordinary family lived up to the ideal of a question seeking an explanation. It is also remarkable that we neither expect nor demand that most of our questions be answered. This indicates that questions offer something else, something greater than answers and knowledge offer. This special something is where things get really interesting.

Perhaps it's not our upbringing, teaching and education that determine what questions we hear and don't hear. Perhaps it's not these elements that draw the boundaries between 'the same' and 'something else'. Perhaps what does this is the question itself.

THE STRUCTURE OF THE QUESTION

THE ART OF ASKING

Imagine two people reminiscing in a sun-speckled garden. Neither one says a word, but either can say anything, about any topic, at any time. So far, all options are open.

Then one turns to the other and asks: "How's the new puppy?" This radically narrows down the options. Any response about birdsong, sunlit rooftops, or the hilarious time you had at your sister's yesterday would sound worse than misplaced. It would sound deranged. The moment one person poses a question to another, a framework is set that can no longer be broken.

So too for Eve in the Garden of Eden. The moment she was given the opportunity to answer for herself, Eve became subjected to the urgency of knowing good and evil. The same in Plato's dialogues, where the questions from Socrates not only open a passage through which we can come to know the doctrine of *ideas*, but also close off any understanding the interlocutor may have had in advance – and any other possible understanding.

The point of Plato's dialogues is that Socrates' questions lead to a particular way of thinking and understanding the world that is better than any other understanding.

That makes it quite fitting to have Socrates ask all the questions: The passage from question to answer – provided one manages to ask the right questions – is a passage from a bad state to a good state.

So what about the rest of us? Those who can't manage to ask the 'right' questions? Or who get 140 'wrong' questions before stumbling into a 'right' one – which we then fail to answer? What options and passages do we open and close when posing a question? And is it even correct to say that *we are posing* the question? Perhaps *the question is posing* us: positioning us in a particular understanding of, and approach to, the object of the question?

The openness we usually associate with questions is a limited openness, as the German philosopher Bernhard Waldenfels puts it, in that the question *draws a boundary*. However many questions, and whatever the topic and wording, the question's paradoxical essence means there is always something we are *not* asking about.

The question now is about the nature of the unasked element. More precisely: If something *has not* been called into question, does that mean we *cannot* question it?

Derrida explains that the reason Western philosophers haven't called the question into question is this element in every question that we *cannot* question. Now, here's the catch: The 'something' we cannot call into question is not a specific, particular thing. Truly, there is nothing in this world we cannot enquire about or criticise, not even questions themselves. But the catch is that *whenever* we ask about something specific, there's always something else we cannot ask about, and that 'something else' *is* a specific, particular thing.

Few other phenomena are as important to our culture and daily lives as questions are, and yet they receive so little attention. This raises the question: Why questions, in particular? It also raises the question: How can such a thing – the question – escape critical attention, not only over a long historical period, but also throughout each individual's life? And the latter must certainly be the case if we can ask some 320 questions a day without even being aware of it.

When we ask a question, we always have prior knowledge of what we are asking about, because we are always part of a historical tradition. The historical tradition is what makes us ask as we do. Or alternatively: Our norms set the framework for our curiosity. Therefore the assumptions made historically about questions hold great authority in the culture within which they are transmitted – such authority that no one notices them, or questions them. Therefore these assumptions could be different, and presumably are in other cultures and societies.

Precisely because some culturally defined norms are always implied when we ask a question, Derrida suggests that we are better served by not asking. He also says, however, that there is a *technology* associated with asking questions; a technology that casts the object being asked about in a particular light. If the nature of the question is essentially technical, then like every other technology it consists of elements that can be taken apart and reassembled. This must mean, for one thing, that there are multiple ways of fitting the elements together and, for another, that 'something' other than the question itself is causing the elements to be assembled as they are.

THE QUESTIONER IS IN CHARGE

Heidegger holds that every question consists of three 'structural moments'. Let's take the liberty of listing these as: 1) *that which is asked about*, the question's 'what', meaning the object or matter enquired about; 2) *the one interrogated*, the question's 'who', meaning the person presumed able to answer; and 3) *that which is to be found out by asking*, the question's 'why' or 'target', meaning its framework for how the object can manifest itself.

To exemplify these three structural moments at work, imagine this scenario: Jack, now grown up and long past fetching pails of water, very sensibly asks Jill, "Would you like to go see a film with me?" *That which is asked about* is a trip to the cinema, the matter Jack and Jill are talking about (1). This happens to be an activity, but it could just as well be an object, or a person, or any other phenomenon. *The one interrogated* is Jill, the receiver of the question, who is expected to answer (2). And *that which is to be found out by asking* is the choice Jill has between accepting and declining Jack's invitation to the cinema (3), so the question's purpose, its target, is to induce Jill to make a decision.

The first two structural moments may give the impression that the question primarily concerns a trip to the cinema (1) and Jill (2), but something interesting happens when we reach the third structural moment.

Both the cinema trip and Jill now reveal themselves in the light of something different from them both: the choice between 'yes' and 'no' (3). According to Heidegger, this choice, which is the question's target – and hence the framework within which it can and should be understood – is determined neither by the trip nor by Jill, but by Jack.

It's the questioner who uses the question to target something definite in *that which is asked about* (1). Hence, the questioner also frames the trip to the cinema in a certain way. Suppose, instead, Jack had asked: "What do you think of going to the cinema?" Then *the one interrogated* (2) would still be Jill. However, *that which is to be found out by asking* (3) would no longer be a choice between 'yes' and 'no', but Jill's opinion of the cinema. Put simply: Changing structural moment 3 also changes how 1 – here, the cinema – manifests itself.

The trip to the cinema would no longer manifest itself as a potential participatory activity. It would now manifest as a phenomenon on which one can hold a variety of opinions. Had Jack asked "When did you last go to the cinema?" or "Do you know where the cinema is?", then the cinema would manifest in new, alternative phenomenal forms now related to time and place.

If we stick with our example, Heidegger's point is that neither structural moment 1 (the cinema) nor 2 (Jill) sets the framework for how the cinema manifests itself. Rather, the framework is set by structural moment 3 (*that which is to be found out by asking*) and therefore by the one seeking an answer: the questioner (Jack).

In Heidegger's reasoning, the precondition for a thing's ability to manifest or reveal itself lies beyond the three structural moments themselves. It lies in the questioning way of being that causes the three structural moments to position themselves as they do. What is this preconditional 'something' beyond the three structural moments that determines their positioning? The questioner! In our scenario, it is Jack. He poses the question, thereby demarcating the question's *quest*. By identifying this quest, the questioner also delimits how *that which is asked about* (1) can manifest, and how *the one interrogated* (2) can react.

Hence, Heidegger's brief analysis of the question's three structural moments is merely a prelude to a much deeper analysis of humanity's enquiring mode of Being. Here, the question of the question is replaced by the question of what it is to be human. Thus, Heidegger joins the long Western tradition of placing humans and our responsibility as human beings at the centre of the world.

All the same, his analysis of the question's three structural moments also opens other options. Taken together, my observation that they can behave very differently in different examples and my conclusion that something beyond the three structural moments determines how they position themselves raise this question: *Must* the questioner be the determining factor? Couldn't it just as easily be something else – like which language the questioner must use to ask a question?

QUESTIONS AND LANGUAGE

THE QUESTIONING MIND

Back to the question of which came first: language or the question. Or perhaps not? Earlier we asked about the first question in the history of humankind. Now we are asking: How does our nature as linguistic beings influence our way of asking, and thereby of thinking, being, and knowing?

Our fictitious friend 'Asking Andy' reminded us that we are bounded in space and time. Even his seemingly nonsensical questions about why a boot or a dresser can't talk illustrate that we are defined by certain basic conditions. Like our limited extension in space and time, our use of language gives expression to the fact that we are finite beings.

As children we don't come to know the world through language in general, but through a particular language. We are not language-minded creatures. We are Danish-minded, German-minded, Chinese-minded, and so forth, depending on our mother tongue. And given that, say, Danish and Russian grammar differ immensely, just imagine how differently we might enquire into the world, ourselves and each other in different linguistic cultures.

A RUSSIAN-MINDED QUESTION?

"Instead of i if write y what will be?" This English approximation reflects just one of numerous questions that was recorded for my research while I was observing a Russian private-school classroom session in Denmark. The question was interpreted verbatim, with no regard for English grammar, which is presumably why it looks like gibberish. What is the teacher asking about? Who or what is the subject of the question? And is there an object?

The answers are not obvious from the question, but perhaps that's not necessary. The pupil wasn't confused at all about how she was expected to answer, and neither teacher nor pupil appeared to find anything missing. Yet from a Danish perspective, 'something' is precisely what's missing here. We never find out what the question's subject matter was, because *that which is asked about* (structural moment 1) is never mentioned.

In order to answer the question, *the one interrogated* (2) must not only be familiar with but actually inhabit the context in which the question is asked. The pupil certainly fulfils these criteria. Teacher and pupil both know the question's aim is to prompt the pupil to demonstrate that she knows and has understood *that which is asked about* (1) so well that the teacher doesn't need to name it explicitly.

Actually, there is *not* 'something' missing. On the contrary, the requested object or matter (1) is so strongly present as to be self-evident. For Heidegger, it is the questioner who sets up the framework for the question. Not so here. Here the framework is set by the object or matter that teacher and pupils share, and precisely because they share it, it can safely be left out of the question.

This could indicate that Russians economise differently with their grammar than Danish- and English-speaking people do, and that such economising – here as omissions – can help us understand the difference between different question cultures.

Isn't that precisely what omissions are: an expression of the fact that in every language culture, some things are taken so much for granted that no one finds it necessary to ask about them? And doesn't it make a difference whether what is 'forgotten' is the human being or the subject-matter; a difference with ramifications for how we enquire about and relate to everything else?

The modest format of this book prevents me from answering these questions here. Suffice it to say, not all cultures of questioning give as much weight to the human questioner and responder as we do in the West. This shows in language structure, and also in our concrete handling and use of questions.

I have observed Danish schoolteachers, and for some, 80% of the questions they ask in class are of a practical or social nature: "Where is your pencil case?" or "Don't you understand that you need to stop being so noisy?" This means only 20% are thematic questions such as: "How do you spell 'he'?" or "Is 'ø' a consonant or a vowel?"

This hardly ever happens in Russian or Chinese class-rooms, where the teachers I've observed often ask up to

90% thematic questions, just 10% practical questions and virtually no social questions at all.

These morsels of information surely tell us a good deal about Danish, Chinese and Russian teaching cultures, but that is not my point here. My point is to suggest that the structure of a language provides guidelines for how users of that language do and do not ask questions.

BETWEEN NORMS AND CURIOSITY

The last time I read *Spørge-Jørgen* to my daughter, she fully appreciated that the boy's parents were fed up. 'Asking Andy' and his questions were "silly", she said. When I asked her how you can tell a silly question from a good one, she gave me a look of equal parts surprise and reproach: "Don't you know?"

A silly question, I came to learn, is one either *everyone* can answer, or *no one* can. Everyone can say why Asking Andy's backside is sore: He's been spanked. No one can say why people don't have eyes in the back of their heads: That's something no one knows, or can know.

So my daughter rejects norms and curiosity as guiding principles for good questions. Rather, she rejects them when they act in isolation. Judging by the questions she herself asks, she is clearly keen on curiosity when it calls our norms into question.

To uncover my daughter's definition of a good question, we must derive it from her definition of a silly question: A good question is one that *some people* can answer, while *other people* can't. The question is, then:

What determines whether a person can or can't answer a question? Well, the answer lies in the question.

Let's revisit Gadamer, who also says the question's essence is to challenge humans, placing them on a knife's edge between 'yes' and 'no'. But the question does much more than this. That's why, even in a small book, we cannot treat the question without also addressing good and evil, knowledge and cognition, freedom, will, responsibility, and a host of other weighty themes.

Questions play an important role in our understanding of and approach to all these themes. This could be taken to indicate that the question is the key to understanding the inner logic that links all major themes in the history of Western philosophy. If that is so, then the question's essence is not to challenge humankind by placing us on a knife's edge between 'yes' and 'no', but rather to identify – and absorb us into – a particular way of thinking and relating to our surroundings.

If this is the case, then the question is more than just the key to understanding Western philosophy's history and inner logic. Hidden within it, the question also bears the truth about what lines of thinking create meaning, direction and coherence in other traditions and cultures than our own. In short, there is actually good reason to call questions into question.